CANOPIES iN tHE Clouds

EARTH'S RAIN FORESTS

by Ellen Hopkins

Perfection Learning®

Cover design and inside layout: Michelle J. Glass

About the Author

Ellen Hopkins lives with her family, four dogs, two cats, and three tanks of fish near Carson City, Nevada. A California native, Ellen moved to the Sierra Nevada to ski. While writing for a Lake Tahoe newspaper, she discovered many exciting things and fascinating people.

Cover images: ©CORBIS royalty-free—fern leaves; Hemera Studio—frog and sloth; Corel Professional Photos—boy; ArtToday—flower

Image credits: ©Lightstone/CORBIS p. 35 (top); ©Daniel Lainé/CORBIS p. 35 (bottom); ©Charles & Josette Lenars/CORBIS p. 36; ©Wolfgang Kaehler/CORBIS p. 39; ©Paul Almasy/CORBIS p. 41; ©CORBIS p. 42; ©Richard Bickel/CORBIS p. 45; ©Catherine Karnow/CORBIS p. 61

ArtToday (some images copyright www.arttoday.com) pp. 5, 7, 8, 9 (bottom), 10, 13, 14, 16 (top), 18, 19, 21, 22, 24 (top), 25 (right), 26, 27, 28, 29 (bottom), 30, 31, 38, 47, 50, 51, 53, 54, 55, 56, 57, 58, 59, 60; Hemera Studio pp. 20, 24 (bottom); ©CORBIS royalty-free p. 23; Corel Professional Photos pp. 4, 9 (top), 17, 25 (left), 29 (top), 32, 33, 34, 49, 63

Table of Contents

Introduction

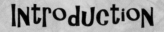

CANOPIES IN THE Clouds

EARTH'S RAIN FORESTS

Imagine cruising though treetops 200 feet above the ground! Millions of **species** of animals do just that. A whole community of insects, birds, reptiles, and mammals lives in the **canopies** in the clouds. This rich community is one of Earth's oldest **ecosystems**. It hasn't changed much in 100 million years!

The wet, green canopies of Earth's rain forests buzz, growl, and chatter with life. Scientists haven't even been able to identify all the creatures. Let alone study them!

Man cuts, clears, and burns the rain forest. So **flora** and **fauna** disappear daily.

Are the rain forests in danger? Some scientists don't think so. Others are quite sure they are. If so, what does that mean to the rest of the world? What does that mean to *us*?

Let's explore.

CHAPTER 1

Wet Makes Wild

Flying over the Amazon rain forest in South America, we can't tell much about what's below. It looks like a puffy green carpet. Are all those trees the same? What, exactly, *is* a rain forest? Is it a hot, wet, and steamy jungle? Is it overgrown with vines and bushes? Is it full of creepy, crawly critters? Well, yes and no.

Most rain forests grow where it's very warm—like here in South America. These tropical forests grow in a wide band within 23.5 degrees north or south of the **equator**. That's why these are also called equatorial rain forests. We'll explore other types of rain forests later.

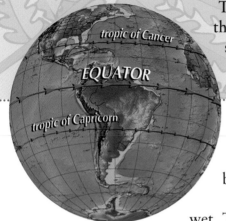

The region north of the equator is the tropic of Cancer. The area south is the tropic of Capricorn. That is where the word *tropical* came from.

Daytime temperatures in this equatorial belt average 86°F. At night, they rarely fall below 70°.

All rain forests are certainly wet. The average rainfall in these areas is 80 inches or more every year. Compare that to Los Angeles. The rainfall in that coastal city averages only 12 inches per year. Just east of Los Angeles is Death Valley. This desert area gets less than 2 inches of rain each year.

Even more incredible, some areas in the tropics get as much as 300 inches every year. That's almost an inch every day! No wonder we call them rain forests!

Combine all that rain with warm temperatures. Then you can also call rain forests steamy. As rainwater **evaporates,** the mist rises back into the sky. There it cools and becomes rain again. Half of all rainfall here is due to this recycling.

As we climb from our plane, the **humid** air feels heavy against our skin. Our hair frizzes with moisture. We start to sweat. But we're excited. Our rain forest adventure is about to begin.

At first, we'll travel upriver by boat. Trees, shrubs, and flowers grow thickly along the riverbank. From a distance, it looks somewhat like the Mississippi River area in the Deep South.

But up close, most of these plants are different from the ones along the Mississippi. And what variety!

6

Some 300 kinds of trees grow on every 25 acres of land. Most North American forests have only 15 types of trees on the same amount of land. But that's not all. On a single acre of rain forest, as many as 100 different trees might be found!

Think how amazed early rain forest explorers must have been. They traveled by boat like we are. Noticing the tangled riverbank undergrowth, they called rain forests *jungles*. But that description doesn't really fit.

The word *jungle* comes from *jangal*. In Hindi, it means a "wild, waterless place." Rain forests might be wild. But waterless? No way! In English, *jungle* means "vegetation so thick you have to chop through it." Remember Tarzan movies? Guys had to whack their way through with **machetes**.

But don't worry. As we set off on foot, we won't have to do much chopping. Vegetation grows thickly along riverbanks because it gets lots of sunlight. Inside the rain forest, it's very different. Sunlight cannot reach beneath the broad, leafy canopies. So underbrush is scanty.

Bamboo

Our guide points out a wide trail used by native hunters. We slip through the curtain of growth along the riverbank.

It takes a few seconds for our eyes to adjust to the dim light. When they do, we feel like tiny Alices in a dark green Wonderland.

Did we shrink? Everything is so big! Bamboo stretches 100 feet into the air. But if it's trying to outgrow the trees, it's got a long way to go.

Philodendrons

Monstrous evergreens grow in layers. The tallest reach almost 200 feet. Tree trunks look like skyscrapers. They're covered with vines as big around as a man's body. These vines drip colorful flowers the size of dinner plates.

Near the ground, we find shade-loving plants. Light green ferns, dark green palms, and white-striped dumb canes are everywhere. Philodendrons (fil-oh-DEN-druns) weave their way up the trunks. At home, people grow these as houseplants.

Most of these **understory** plants never grow taller than 15 feet. They simply don't get enough sunlight. But their oversized leaves soak up as much as they can.

These huge leaves also catch moisture that trickles down from above. Because down here, things don't get very wet.

Way above our heads, we hear rain drumming on the leaves. But we can't feel it. The gigantic trees and plants of the upper canopies shelter us like umbrellas.

Rain forests have from one to three canopies of flat-crowned trees. These giants rise between 100 and 165 feet above the ground.

Hey, check it out! It's a "Mega-tree"! This 250-foot monster is an emergent. That means it's so tall, it rises above the other trees. We can't even see their rounded tops poking through the upper canopy. As we wander through the rain forest, we'll find maybe one emergent per acre.

Fungi

We study Mega-tree's straight, white trunk. It's covered with **fungi** and **lichens** (LIE-kuns). At the bottom, the trunk looks swollen. The big knobs, or *buttresses*, brace the top-heavy emergent in the ground as it sways in heavy winds.

Rain forest trees also get support from woody vines called *lianas* (lee-AH-nahz). Lianas climb from the ground to the canopy. They attach themselves to trees with "hooks" as they grow. Some reach 800 feet long.

As lianas grow, they wind around and between trees. Often, they loop through hundreds of yards of forest to form a net of vines.

Lichens

Epiphytes (EH-pi-fites) grow high in the canopy. These include orchids, ferns, bromeliads (bro-MEE-lee-ads), and even cacti. Up there, sunlight bathes these epiphytes. They flourish.

Bromeliads

Sometimes epiphytes get very heavy and break the branches on which they're growing. To fight back, many trees have evolved. They now have smooth bark or nasty **toxins** that epiphytes don't like.

Most plants take water and nutrients from the soil with their roots. Epiphytes are also called *air plants* because they don't root in soil. Instead, their roots wrap around branches, leaves, and trunks.

Epiphytes have thick, waxy leaves that collect water. They feed on dust, animal droppings, and other litter that falls into the trees. But they don't feed off the trees themselves.

Plants that attach themselves to trees and take nutrients from their hosts are called *parasites*. One parasite, a strangler, starts as an epiphyte. Then it sends long roots down the tree trunk and into the soil. In the soil, the roots choke, or strangle, the tree's roots. Eventually, the host tree dies.

Treetops. Liana nets. Epiphytes. All that growth above creates a lot of shade below. Only 2 to 5 percent of the sun's rays can filter through. Most plants need sunlight for energy. But in the gloom on the ground, we find the shade-loving understory plants. Otherwise, the forest floor is bare except for saplings and seedlings.

See how those baby trees struggle toward the gentle light. All their food energy is stored in their seeds. Once that's used up, they grow very slowly.

Most saplings will never mature and become part of the upper canopies. Their only chance of survival is if gaps appear in the cover above them. Gaps form when bigger trees die of old age or fall prey to wind and weather.

If a hole in the canopy opens, little trees race toward the sun. The fastest reach the upper levels first. It takes them about 100 years to reach the upper canopy. Slower, stronger trees take 200 to 500 years to get that tall. But in time, they replace the fast-growing trees.

Meanwhile, the extra light also helps other seeds sprout. New seedlings appear and grow slowly until given the chance to race toward the sun. This endless cycle keeps the rain forest alive.

High overhead, wind whistles through the treetops. Our guide tells us the wind often reaches **gale force**. But here on the forest floor, the air is still, warm, and, of course, moist.

Humidity in the rain forest stays about 70 percent. Such warm, wet conditions favor fungi. Fungi break down, or rot, fallen trees and leaves.

We don't see much **debris** around us because it is quickly "eaten." Rotting debris releases **nutrients** for which rain forest plants hunger. They don't get much food from the soil.

Rain forest soil is poor because all the rain washes **mineral** nutrients away. So the plants here compete as fiercely for minerals as for light. Many form "partnerships" with the fungi that grow on their roots.

Fungi take energy from the trees they grow on. In return, they give the trees phosphorus and other nutrients that they manage to find in the soil. This sort of give-and-take relationship is **symbiotic** (sim-bee-AH-tik).

We see many of these relationships in the rain forest. For instance, trees send out roots to search for food. Since most nutrients are found near the surface, the roots don't go very deep. Shallow rooting means less support, which means more trees fall during storms. That opens up more holes in the canopies, giving more sunlight to struggling seedlings.

These cycles have continued for millions of years. Let's travel back and take a look.

CHAPTER 2

WHere iN the WorLd?

Earth—100 million years ago. Scientists call this time the Cretaceous (cree-TAY-shus) period.

The world was much different then. It was warmer and wetter. Crocodiles slogged through swamps in Alaska. Pterodactyls flew over groves of Antarctic palm trees. Dinosaurs roamed the **conifer** forests growing near the North Pole.

Then new kinds of plants appeared. Flowering plants. The hot, humid climate suited them perfectly. They spread across the globe. Soon, the entire planet looked like a tropical rain forest.

The Cretaceous period was also a highly volcanic time. Over and over, lava and ash spit up into the **atmosphere**. After millions of years of eruptions, a lot of ash collected. The earth couldn't get as much sunshine. The planet cooled, especially near the North and South Poles, which get less sunshine anyway. In those places, great sheets of ice called *glaciers* formed.

Cooler temperatures meant less water evaporated. That meant less rainfall. So the world became drier.

Many trees and plants died. Others had to evolve to survive. Some tropical plants didn't evolve. So they could only grow near the equator where temperatures stayed warm.

Since then, tropical plants have changed only a little. In many ways, tropical forests look the same as they did in the Cretaceous period. Minus pterodactyls and dinosaurs, of course.

Today, tropical rain forests cover about 7 percent of the earth's surface. They grow in three main regions of the world. The largest is the American rain forest. It begins in eastern Mexico and runs through Central America to northern South America. The biggest chunk of that area is the Amazon basin.

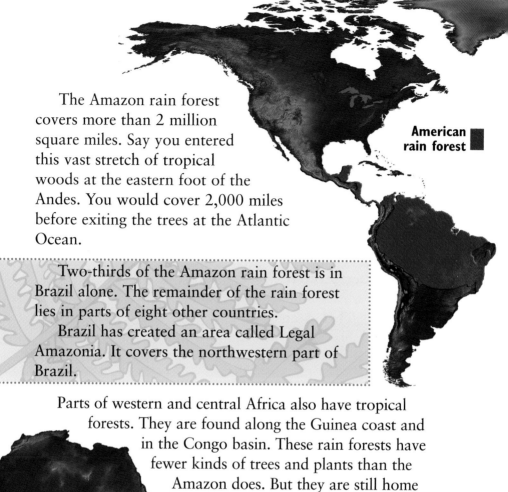

The Amazon rain forest covers more than 2 million square miles. Say you entered this vast stretch of tropical woods at the eastern foot of the Andes. You would cover 2,000 miles before exiting the trees at the Atlantic Ocean.

American rain forest

Two-thirds of the Amazon rain forest is in Brazil alone. The remainder of the rain forest lies in parts of eight other countries.

Brazil has created an area called Legal Amazonia. It covers the northwestern part of Brazil.

Parts of western and central Africa also have tropical forests. They are found along the Guinea coast and in the Congo basin. These rain forests have fewer kinds of trees and plants than the Amazon does. But they are still home to millions of flora and fauna.

Central African rain forest

The third tropical rain forest region is known as the Malesian subkingdom. This area stretches from Myanmar (once called Burma) to Fiji. It includes Thailand, the Philippines, Indonesia, and parts of Indochina and Australia.

Malesian subkingdom

Even within the equatorial rain forests, conditions vary. Let's explore the reasons.

We climb into a helicopter for a South American whirlybird tour. From the heart of the Amazon, our pilot flies west.

As we hover over Colombia and northern Ecuador, we see a huge tract of rain forest. This area is called the Choco. It is the wettest region of the world.

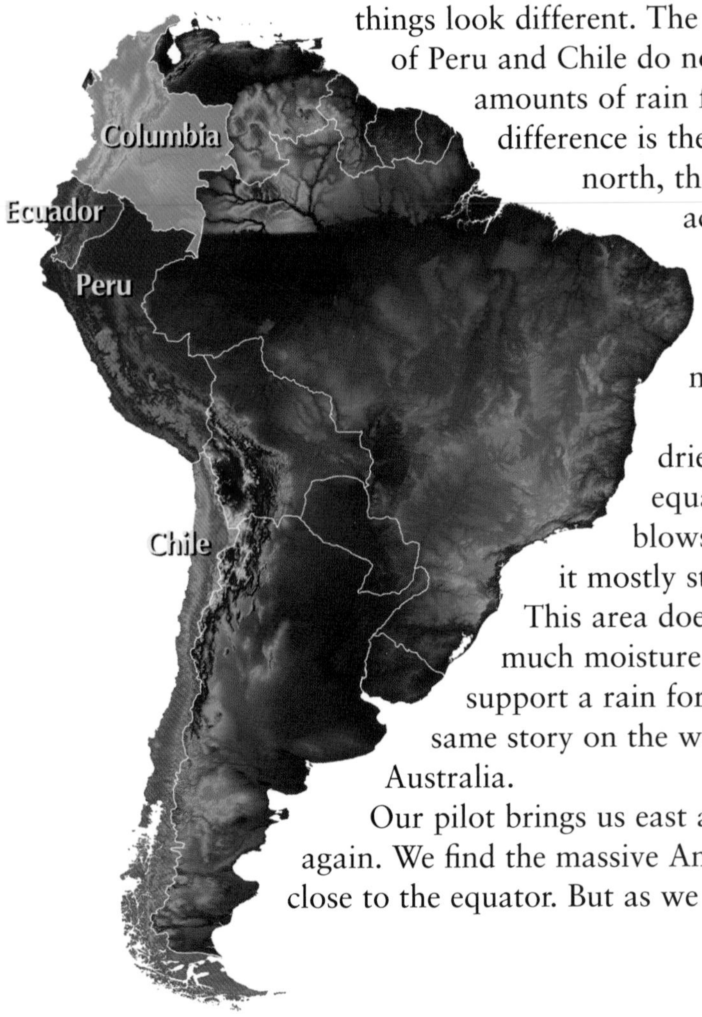

Now we turn south. Below the equator, things look different. The western coasts of Peru and Chile do not have large amounts of rain forest. The difference is the wind. To the north, the wind blows across the ocean and onto the shore. It brings moisture with it. Things are drier south of the equator. The wind blows here too. But it mostly stays offshore. This area does not get as much moisture and cannot support a rain forest. It's the same story on the western coast of Australia.

Our pilot brings us east and north again. We find the massive Amazon forest close to the equator. But as we move farther

Columbia

Ecuador

Peru

Chile

16

north, the forest begins to look different. We see some **deciduous** trees. The upper canopy is more open and there are fewer lianas.

This kind of forest is called **subtropical** rain forest. The difference here is temperature. As you move away from the equator, the climate is less constant. Sometimes it's hot, sometimes it's cooler. That means seasons.

Where seasonal differences are small, you'll find subtropical forests. Besides here in Brazil, they're found in the West Indies, Southeast Asia, and eastern Madagascar.

In places with long, dry seasons, we find monsoon rain forests. A monsoon is a wind that changes directions with the seasons. When it blows one way, everything gets very wet. When it switches directions, the area becomes parched with **drought**.

Monsoon rain forest

Monsoon rain forests have mostly deciduous trees. They don't grow as tall as tropical trees because they don't get as much water. Annual rainfall is less than 50 inches. The dry season might last four or five months. During this time, the trees lose their leaves. That lets more light into the understory. So monsoon forests often have dense undergrowth.

This kind of rain forest is widespread in Myanmar, Cambodia, Thailand, Indonesia, and West Africa. But we don't see much monsoon rain forest in South America.

Tropical and subtropical coasts are sometimes bordered by mangrove forests. These special forests are not true rain forests, but they often form along their edges.

Mangrove forest

What makes mangrove forests special? Sometimes the ocean floods them. It leaves the ground soggy and very salty. Few trees can handle that. Those that can rarely reach 30 feet. In the whole world, there are fewer than 50 kinds of trees in mangrove forests.

One factor that affects rain forest growth is **elevation**. As you climb a tropical mountain, you lose about one degree of temperature every 350 feet. Our helicopter follows tropical rain forest to the eastern slope of the Andes. As we fly higher up the face of the mountain, the forest changes again.

At about 4,500 feet, we start to see lower **montane** rain forests. There's plenty of rainfall here. But the cool temperatures do not support as many tropical plants. We do see ferns and bamboo. In sunny clearings, they grow very large.

The upper canopy of trees reaches 100 feet in places. But it's shorter than the rain forest at the foot of the mountain.

Above 9,500 feet, upper montane rain forest takes over. This eerie land is also known as cloud forest because a bank of misty clouds lingers here. The fog comes from moist air currents that bounce off the mountains and rise.

Here, the sun rarely shines. We see more **temperate** trees such as oaks. Their gnarled trunks climb barely 60 feet high. Their branches are covered with mosses and liverworts, which thrive in the cool mist. Their leaves drip constantly because these tricky trees grab moisture from the clouds. The process is called *cloud stripping*.

We go even higher up the mountain. At around 11,500 feet, we find *really* weird-looking trees poking through the clouds. They're stunted, barely 15 feet tall. Their trunks twist like corkscrews because this elflike woodland is tortured by wind and weather.

Above the tree line of this "subalpine forest," we see little except fern meadows and tundra. Finally, there is snow. Yes, snow even here at the equator. You just have to gain enough elevation. Montane rain forests can also be found in Central Africa and Indonesia.

Tundra

Finally, we find temperate rain forests. These grow mainly along the Pacific coast of North America. Winters are mild. Rainfall is high. But there are only a few kinds of trees. Most are cone-bearing evergreens like redwoods and cedars. Laurels fill in the gaps.

Besides the highly **diverse** plant life, rain forests are home to countless animals and insects. Let's return to our Amazon wildlife trail and have a look.

CaNopy Critters

Entering the evergreen gloom, one thing immediately strikes us. The quiet. We know millions of creatures dart, fly, and glide above our heads. But all we hear is the whine of mosquitoes and the twitter of songbirds. Where are all the animals we've heard so much about?

They're all around us. We just have to look closer. Some are hiding. Others blend so well with their surroundings that we don't even notice them. Still others are so high up in the canopy we'd have to climb to find them.

Lots of critters live here. And most of them live in the trees.

Four square miles of rain forest may have 400 kinds of birds, 125 types of mammals, 100 species of reptiles, and 60 kinds of amphibians.

What about insects? Good luck counting all those! A Smithsonian researcher counted 405 different insects on one rain forest tree.

The Amazon alone is home to more than 10 million animal species. Most, of course, are bugs.

As our eyes get used to the pale light, we begin to see life. A flash of electric blue catches our attention. It's a beautiful blue morpho butterfly. It's one of over 150 colorful butterflies. Many are quite unusual. The blue morpho has a wingspan of 6 inches.

Blue morpho butterfly

The Queen Alexandra birdwing butterfly has a 12-inch wingspan. So does the atlas moth. Either could stretch its wings wide enough to cover a phone book! But both of these butterflies live in the rain forests on Papua New Guinea.

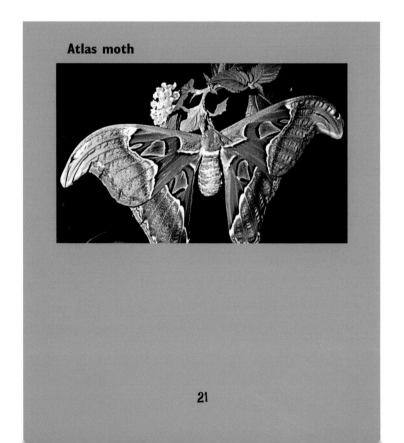

Atlas moth

Like the trees, many rain forest insects are supersized. For instance, the Borneo walking stick grows 13 inches long. Good thing they're only homely, not poisonous! That goes for the 11-inch millipedes that march through the muck. Slugs as big as bananas slither behind them. Luckily all they want to chomp is rain forest vegetation.

Millipede

In fact, the bugs that bother us most aren't giant. They are tiny. Mosquitoes by the millions buzz beneath the canopies. These nasty little creatures don't take much blood when they bite. But they leave their saliva behind.

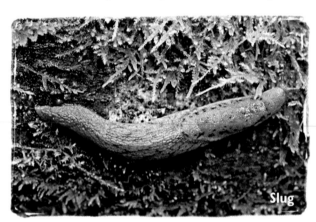

Slug

Mosquito saliva is what makes you itch. It also carries worm **parasites** and diseases like malaria and yellow fever. Malaria kills several million people every year. Good thing we've got plenty of insect **repellent** and mosquito netting.

The netting is extra-fine because a larger weave would let the biting midges (mi-GEES) through. These tiny flies are only $1/25$ of an inch long. You can barely see them, which explains their nickname, "no-see-ums."

But you sure can feel them, especially when a swarm attacks. Like mosquitoes, midges take blood and sometimes leave parasites behind. And their bite can itch for days.

We come to a slow-moving stream. As you might expect in such a watery world, the rain forest is full of streams. They crisscross the forest floor. Many lead straight to the Amazon River. All are very alive.

Some 1,500 kinds of fish have been found in the Amazon waterway system. Many more have not yet been identified. Just beneath the surface, we find several. Giant catfish and pirarucu (pi-RAH-ri-koo) are among the largest freshwater fish in the world. Most of these fish travel in large schools. They **migrate** with the seasons.

Pirarucu

Also in the stream are the small, but deadly, piranha. We cross the stream carefully, hoping to avoid those.

Piranha usually feed on other fish. But they didn't get a fierce reputation for nothing. If they smell blood, a school of piranha will attack a large animal. Yes, even a person. They can strip a carcass to bones in minutes.

Piranha are more dangerous during low-water periods. There are fewer meals to choose from then. So people are more likely to be attacked.

Other hazards also hang out near water sources. See that "log"? Don't sit on it! You wouldn't want to make an anaconda mad.

Anaconda

Anacondas, or water boas, are the largest snakes in the world. The biggest one ever caught was 37½ feet long and weighed over 500 pounds.

Snakes are not common in the rain forest. At least not on the ground. Snakes feed mostly on rodents or birds. Since the forest floor isn't cluttered, you don't find a lot of small rodents. So most rain forest snakes are up high, hunting birds.

Besides the rare anaconda, you might come across a boa or python. Don't worry. These snakes swallow their food whole. They never attack prey too large for them to eat. You probably won't be on the menu.

Yes, there are venomous snakes. But they're not waiting to bite you. Few will attack unless you bother them. Most are patient—even lazy.

One researcher studied rain forest fauna for 25 years. In all that time, he never even came close to getting a snakebite.

Crocodiles and caimans never bothered him, either. Both live here in the Amazon. They're hunted for their skins. Crocodiles are threatened by overhunting.

Caiman

Caimans are South American reptiles similar to alligators. But they look more like crocodiles.

Crocodile

Crocodiles have walked on Earth for some 200 million years. They are the last of the "ruling reptiles." They have survived when bigger, meaner dinosaurs all died.

The Amazon is home to the American crocodile. This croc would rather swim than fight. But you wouldn't want to run into his African cousin.

And India's saltwater crocs sometimes measure 20 feet long. They are some of the nastiest animals in the world. They have been known to attack canoes 100 miles out to sea!

Whoa! Did you see that spotted flash? It dashed up that tree. That ocelot wanted nothing to do with humans stalking his habitat. Both he and his rain forest cousin, the jaguar, are hunted for their beautiful pelts. People are their only enemies here. We're lucky to have glimpsed this one.

O—o—o! A—a—a! Harumph! Hear that? The ocelot's flight has excited a group of howler monkeys. The loud-mouthed howlers are the Western Hemisphere's largest monkeys.

Way up there, above our heads, live a large assortment of **primates.** Woolly monkeys, capuchin monkeys, spider monkeys, and tiny marmosets live in the trees. All are hunted for food.

Bullets have now replaced poison darts as weapons. As a result, many rain forest primates could soon become endangered.

Spider monkey

Marmoset monkey

Other animals are captured to sell to pet lovers around the world. They include many tropical fish and birds like parrots. You've probably admired the birds' brilliant **plumage** at pet stores.

Here in the rain forest, birds fill the canopies like colorful flowers. Eagles and hawks dive-bomb giant flocks of parakeets. Woodpeckers whittle tree trunks. Herons, spoonbills, and scarlet ibis liven up waterways. At dusk, macaws fill feeding grounds, and toucans cry in the treetops.

We'll want to be out of the forest by then. As night falls, bats by the thousands take to the air. Among them are vampire bats, hungry for blood.

Most vampire bats feed on mammals, especially livestock like horses and cattle. They bite through the skin, then lap blood from the wound. They don't take much, maybe an ounce in one feeding. But the wounds often become infected. And some bats carry rabies.

Humans are mammals and are sometimes bitten. But they aren't regular vampire bat targets.

As we turn and head back to the boat, we keep our eyes peeled. If we're lucky, we might spot a coati, an anteater, or a capybara—the largest rodent in the world.

Capybara

Mammals, fish, reptiles, amphibians, and insects all play important roles in the lives of rain forest trees and plants. Let's take a closer look at these symbiotic partnerships.

CHAPTER 4

Canopy Partnerships

So why would big, beautiful, healthy trees *need* to form partnerships? There are so many of them, right? Well yes, there are many trees in the rain forest. But there aren't a lot of each kind. With so many different species, there's only room for one or two of each on any given acre. Some kinds of trees are only found in certain small areas.

The same is true for the animals that call the rain forest home. Some live only in one little area. A few live only on certain trees or plants. For them to survive, special measures need to be taken.

From the tiniest insect to the grandest mammals, rain forest animals have evolved in unusual ways. Many never leave the treetops. Some have developed **prehensile** tails which they can use like arms to hang on to branches. These include apes and such surprising critters as porcupines and anteaters!

Others have learned to fly. Okay, they don't really fly. They glide. There are "flying" squirrels, lemurs, dragons (lizards), frogs, and tree snakes. Watch out for that flying snake! It can cover 160 feet in a single bound!

Up here, treetop dwellers must find all of their food and water up in the canopies. That's easy as long as it's raining. But what about during dry spells?

Mother Nature has created "water tanks." These are bromeliads. They are closely related to pineapples. Many have thick, overlapping leaves. The leaves form a **reservoir** that holds water, even during dry spells.

Big bromeliads hold a lot of water—as much as 30 gallons. Some creatures drink

Lemur

from them. Others eat the algae that grow in them. A few even live in them, including an **arboreal** crab that lives nowhere else.

Bromeliad

So what do the bromeliads get for their trouble? Ants. That's right, ants. They nest in the bromeliads' roots. As they **forage**, the ants bring back bits of food. They share them with the plants that give them shelter.

Ants, in fact, have formed many rain forest partnerships. Let's climb down out of the canopy and take a peek.

Marching along the rain forest floor, we see an army of leaf-cutter ants. The "highway" they follow is litter-free. They keep it that way, making it easier to carry their heavy cargo. Well, heavy for ants, anyway.

Leaf-cutter ants

These ants gather dime-sized leaves from certain trees. They carry them down the trunks, along their highways, and into their underground nests. But they don't eat the leaves. They chew them into **mulch**. This mulch is used to **cultivate** a special fungus that is the only thing the ants eat. The fungus is found nowhere else on Earth.

Leaf cutters weed their fungus gardens and keep them free of pests. When a new queen leaves the nest, she takes a bit of fungus with her. Her new colony can then start its own fungus garden.

Wait, there's more. Trees take root in leaf-cutter tunnels and keep them from caving in. In return, the ants' special fungus kills some tree diseases. Ants and plants protect each other. This partnership guarantees both will survive.

Careful! Don't touch that tree. Oh, the trumpet tree won't hurt you. But its army of Aztec ants certainly will! Aztec ants make their homes inside the tree's hollow branches. They guard the tree against the creatures that crave its fragrant fruit and tasty leaves. The slightest brush against the trunk will bring a vicious swarm of biting ants.

In return, the tree grows little capsules of glycogen. Glycogen is a starch, usually found in animals. The trumpet tree is the only plant known to produce it.

Glycogen makes up most of the Aztec ants' diet.

The rest of the ants' diet comes from mealybugs or aphids. No, the ants don't eat them. They keep herds of these sap-sucking insects. Then they "milk" a sugary nectar from them. Rain forest ants not only garden but they keep livestock too!

All that is just to protect a single kind of tree. In the same way, little beetles help preserve mimosa trees. Mimosas grow in rain forests to the north of the Amazon. These trees don't live very long—maybe 20 or 25 years unless they're pruned. Pruning makes them live twice as long.

Mimosa tree

Enter mimosa girdler beetles. Females will only nest on mimosa trees. With sharp **mandibles**, they slice into branches and lay their eggs inside. But the eggs won't hatch in live wood. So the beetles cut through the bark to kill the branches. Then the eggs hatch. And the mimosas are pruned.

Rain forest animals also help create baby plants and trees. High in the canopies, the wind **pollinates** flowers. But these breezes cannot reach below the treetops.

Down there, plants must rely on critters to do the pollinating. Insects, birds, and bats all help out. Each has favorite flowers to pollinate.

Hummingbirds and flower-peckers sip nectar from flowers. As they do, they become dusted with pollen. When they move to the next flower, they take pollen with them.

When they
move again,
some of the pollen is
left behind.
Most flowers
pollinated by birds are red.
Birds seem to like that color.
Flowers pollinated by night-flying
moths are usually white or pink. Those
pollinated by insects that fly during the day tend to
be yellow or orange. Bats prefer pale, highly scented
flowers that open at night.

Plants also rely on animals to move their seeds from
place to place. Why is moving seeds important? If seedlings
are spread out, there is less competition for food and light.
Also, some insects feed only on one kind of seed. If those
seeds are all in one place, it is more likely all will be eaten.
A little seed "hide-and-seek" helps plants survive.

Bats and birds pluck ripe fruits from the canopy. They
carry the fruit to their perches, eat the flesh, then drop the
seeds. With any luck, the seeds will root and grow.

Other fruits fall to the ground to ripen. Mammals carry
them off, dropping seeds along the way.

You might expect all that. But did you know that fish
move seeds? Many trees drop fruit that the fish find
yummy. Seasonal flooding carries fish deep into the rain
forest. They find their favorite fruit, dropping seeds as they
swim. When the water **recedes**, the seeds take root.

So where's the partnership? Some of these fruits make the fish's flesh poisonous. Predators look elsewhere for an easy meal and leave the fish alone.

Some caterpillars and butterflies also feed on poisonous plants. They have developed **immunity** to the toxins and can eat the plants. But other creatures cannot eat the butterflies!

Beyond bugs and birds, rain forest partnerships have been formed with the most successful animals of all—people. Let's find out how.

CHAPTER 5

Rain Forest Peoples

Jiwiki hunters

We live in a world of highways and high-rises. So it might be hard to imagine a place where people still travel on foot and sleep in huts.

Many rain forest dwellers do exactly that. The men hunt with blowguns and bows and arrows. The women gather from the forest's bounty and tend their hearths. These **primitive** tribes have survived, almost unchanged, for thousands of years.

Pygmy people

The very first people to live in an area are called *aborigines*. Think original. In Africa, the aboriginal rain forest dwellers were the Pygmies and the Bushmen. The Bushmen have deserted the rain forest in favor of the desert. But many Pygmies still call the rain forest home after 50,000 years.

Are Pygmies really small? Yes! They average only 60 inches, or 5 feet, tall.

Most rain forest natives are small. Their size helps their bodies lose heat faster. That's important where they live. In such a humid climate, sweating is not a good way for the body to cool off. So rain forest dwellers don't sweat very much.

Other groups arrived later and established rain forest communities. Many turned to farming or trading. The Ashanti, for instance, are skilled farmers. And also artists. They weave, make pottery, and manufacture gold and silver jewelry. Their ornaments are sold far from their homeland.

Ashanti group

The Ashanti are a culture group that lives in the African rain forest.

35

Bantu woman

About 2,000 years ago, the Bantu people arrived on the scene. They are now the dominant culture in the African rain forest. The Bantu are farmers. They fight the forest, clearing the land for villages, gardens, and grazing. But rain forest soil is poor. It will only grow crops for six seasons. Then the Bantu must clear more land. It's hard work.

Some Pygmies decided to try farming too. They moved to the Bantu villages and changed their way of life. The Bantu weren't exactly friendly. They looked down on the Pygmies and considered them servants. But they also relied on them to go into the forest and bring back food, especially honey.

Honey is an important food source for the Pygmies. It is not only sweet, but it's high in calories.

The Pygmies have formed a partnership with birds called the greater honeyguides. The greater honeyguides eat beeswax. But they have short bills. The birds can't reach deep inside trees where many beehives are. So when the honeyguides find hives, they lead the Pygmies to them with their raspy chatter.

The Pygmies use smoking branches to calm the bees. Then they chop the trees open to find the hives. The Pygmies take the honey and leave the wax for the little birds' reward.

The Bantu wouldn't go deep into the forest. They were afraid of it. The Pygmies took advantage of that fear. Deep in the forest, they were free from Bantu control. And they liked it that way.

Today, the Pygmies live as they always have. These hunter/gatherers work with the forest, not against it. They do not settle in one place. Instead, they follow game, setting up temporary camps along the way.

The Pygmies move fearlessly through the rain forest. And why not? The forest gives them everything they need—shelter, clothing, food, weapons, and medicines.

"Jungle" dwellers lead simple lives. Shelters, which can be built in a few hours, are simple lean-tos. They are made from branches or animal skins.

Since the weather is always warm, clothing isn't really necessary. Some tribe members wear simple skirts. Others wear nothing but body paint, jewelry, or other adornments.

Meals aren't fancy. And they don't always include meat or fish. Trapping animals is hit-or-miss. And hunting with bows and arrows isn't easy! Rain forest dwellers often make do with fruit and simple bread. The flour comes from acorns, roots, or **tubers**.

Sometimes hunter/gatherers go weeks without meat. Then, after a successful hunt, they have a huge meat-eating feast. Their bodies have learned to store protein and other nutrients.

Rain forest people don't get sick very often. Cancer is rare. Stress, **obesity**, and heart disease are almost unheard of. Certain tribes even have some immunity to malaria and yellow fever.

But when they are needed, primitive medicines are plentiful. A single native village in Thailand, for instance, uses 119 different plants for medicines.

Thailand is one of many regions in the Malesian subkingdom. It's hard to pinpoint the aborigines of this area because the South Pacific tribes have mixed for centuries. We do know humans have wandered that part of the world for at least 40,000 years. And of all the places with rain forests, the Malesian subkingdom has the most different tribes.

Humans first reached the Western Hemisphere some 20,000 years ago. Much debate has centered on how they found their way to Central and South America. Some experts say they came by land. Others say they came by boat. Either way, people reached the Amazon Basin about 15,000 years ago.

Many bands moved into the Amazon rain forest proper. Like the Pygmies of Africa, these bands of people stayed **nomadic**. Other tribes set up villages along riverbanks. These tribal "nations" turned to farming and often traded goods with one another. Their settlements prospered and grew. A few had as many as 2,000 people.

The traveling tribes stayed small. It was easier to move through the forest that way. They developed all kinds of snares, traps, blowguns, harpoons, and arrows. They also discovered plants that provided "drugs" to stun fish or game without poisoning them.

Many rain forest tribes were—and are—warlike. As we wander through the Amazon forest, there are some we'd rather not run into. At least if we want to keep our heads!

The Jivaro (HE-var-oh) Indians live at the foot of the Andes. They are famous for shrinking human heads to the size of oranges.

Headhunting was common until 50 or 60 years ago. Today, the practice is dying out. But it's not quite gone. Besides the Jivaro, some Dayak tribes in Borneo still headhunt. And in Indonesia, the tree-dwelling Asmat people go on headhunting raids.

Jivaro man and child

The Asmats are truly "tree people." Their name comes from the words *Os* for "tree" and *amot* for "man."

Many Asmat tribes live their whole lives in the treetops. They believe trees symbolize people. Roots are feet. Fruits are heads. Trees drop fruit so their seeds will make new trees. And so, the Asmats believe, taking an enemy's head means a new life will be born.

Modern governments have outlawed headhunting and tribal warfare. Many even consider their aboriginal people "non-citizens." Modern churches have sent missionaries to **civilize** the "savages." They believe they're doing the right thing.

But do rain forest people need government? Must they become civilized? Should they give up their culture?

WHEN CULTURES CLASH

Who really owns the rain forest? Governments in cities far from the trees? Settlers, who claim the land with axes and plows? Corporations that claim it with chain saws and bonfires? Or the aborigines, who have lived peacefully beneath the canopies for thousands of years?

The idea of ownership doesn't mean much to forest people. Individuals own weapons and household utensils. But canoes or other objects used by everyone belong to the community. Land belongs to the people using it. When they move on, it belongs to no one.

Sharing is second nature to these people. When a hunt is successful, the game is divided. Crops are also shared. Different tribes often trade goods. And when they gather, they exchange gifts.

Government is strange to them. Each tribe has a chief or leader. He speaks for his people and settles arguments among them.

Each tribe also has a *shaman*, or medicine man. The shaman heals with natural medicines. He also believes in magic.

Both the chief and the shaman are given places of honor in their tribe. But neither has total power over the group.

Native religion is not about worship. It's about living in harmony with nature. The aborigines believe good and evil spirits are active in this world. Nature's spirits will help or hurt, depending on how people behave.

Missionaries are generally treated with respect and friendship. Sometimes they do change the natives' beliefs. Sometimes they realize that they never will. Often, the missionaries leave with more respect for the rain forest people and their beliefs.

Other outsiders are viewed with suspicion by the natives. The reasons go way back. Let's look at some history of the Amazon rain forest.

PEDRO ALVARES CABRAL

In April of 1500, Portuguese explorer Pedro Alvares Cabral was headed for India. The wind blew him too far west, and he landed in Brazil.

There Cabral found beautiful people. They were living simply and in tune with the earth. He compared the place to Eden and the people to Adam and Eve.

At that time, between 6 and 9 million Indians lived in the Amazon Basin. A vast network of villages stretched along the river's **floodplain**. And countless tribes wandered the rain forest proper.

Cabral's glowing descriptions lured many Europeans to the Amazon. They brought plows to cultivate the land. And guns to protect it. They also brought diseases. The natives had no immunity against these new diseases. All in all, the arrival of the Europeans spelled doom for many rain forest cultures.

A century ago, Brazil was home to 230 tribes and over 1 million Indians. Today, 143 tribes remain. Combined, their total population numbers less than 50,000. That number is falling rapidly.

Maybe 1,000 rain forest tribes remain in all the world. Some live so deep in the rain forest that they are still unknown. Every now and again, as highways are built and forest is cleared, new groups are found.

Once contact is made, these primitive people have three choices. The first is to move deeper into the forest and thus avoid further run-ins. But as settlers and developers clear more and more land, that is rarely possible.

A good example is the Txukarramae (ta-hoo-kah-RAH-may) tribe. They live in Brazil on the Xingu River. Many years ago, the area was set aside as a **reservation** for 14 Indian tribes. In 1971, highway construction began in the northern part of the park. It split the Txukarramaes' land in half.

The tribe also split. One band settled near the road. Two years later, highway workers infected them with measles. The local hospital had no medicines to treat the natives. Not even aspirin for the fever. Many of them died.

The second band moved deeper into the forest. The land was officially "theirs." But settlers moved in anyway. In one year, the settlers killed 30 Txukarramaes. The Indians were doing nothing but walking on their own land.

When workers began clearing the forest for a cattle ranch, the Txukarramae chief warned them to quit. He said if they didn't stop, blood would be shed. The workers kept cutting down trees. Two days later, the Txukarramaes killed 11 workers. The Indians were punished by well-armed soldiers.

Afterward, Chief Raoni of the Txukarramaes said, "I think the whites should kill us all right away and take our lands and be done with it."

The second choice for natives is to leave the forest. Some tribes have moved to settled areas and gone to work as laborers. Everything for them has changed. They now eat different foods. They wear more clothing. They live in buildings. Many of these things can make them sick. Their bodies aren't used to these changes.

Even worse, people want to change the way the natives talk, behave, and believe. Think how you'd feel if other people said everything about you was wrong. You'd feel sad. You'd feel bad. After a while, you'd probably feel mad. All of those things have happened to rain forest people too.

The third choice is for the rain forest people to accept modern ways but keep their tribal culture. If they're wise, they can even profit from it. The Kuna Indians of Panama have chosen to live this way.

In 1925, the Kuna rebelled against the government of Panama. They claimed 148,000 acres of land. They were to govern themselves. The Kuna Yala territory was recognized by law in 1938. It is now a national park.

Kuna woman

The Kuna women have long been famous for their *mola* blouses. They still wear them. And now they sell molas around the world. You can even find them on the Internet! Besides blouses, molas are now made into tapestries, handbags, and pot holders. They are so popular that many Kuna women earn more money than their husbands.

Today, tourists visit Kuna Yala. They stay in Kuna lodges. They take Kuna canoe trips, tour the rain forest, and buy molas.

Molas are fabric art pieces. The Kuna women make them by cutting shapes from commercial cloth. They sew the shapes onto the front and back of fabric panels. Details like teeth and hair are then embroidered.

Traditionally, the designs are of native animals like toucans, sharks, or alligators. But as the modern world invades Kuna Yala, new images have appeared. They include military aircraft and beer labels.

Tourist dollars have changed the Kuna a little. These people no longer hunt and gather because they can buy food. They have stores with generators to keep soft drinks cold. They can watch TV at the small tourist lodges, although the picture isn't great!

But in many ways, the Kuna haven't changed. They still practice their own religion and marry within the tribe. They still live simply, in harmony with the rain forest. And they still defend it—with weapons if they have to.

Yes, the Kuna officially own their land. Yet each year, nearly 200,000 acres of rain forest are cut illegally. And that includes Kuna rain forest.

Like Panama, many countries have signed agreements to protect their native peoples. But some **treaties** are broken. Laws are not enforced.

The main problem is the land and the wealth it contains. Papua New Guinea enforces the native peoples' rights to their land. Peru claims to, but allows free access for oil and mineral exploration. In the Philippines, one law guarantees tribes the right to ancestral lands. Another allows the government to take those lands for its own use.

At first glance, tropical forests seem to offer unlimited natural resources. But they are such **fragile** ecosystems! Mining. Logging. Ranching. Road building. All these activities upset the delicate balance that keeps the rain forest—and its creatures—alive.

A Disappearing Act

Worldwide, some 93,000 acres of rain forest fall every day. That averages out to 3,875 acres every hour or 64 acres per second. When forest is lost, partnerships are lost. Species, including humans, disappear. Often forever.

Is the rain forest in danger? In some places, conservation has helped it rebound a little. In others, it has almost vanished.

To explore further, let's take to the air again. But this time, a helicopter won't do. We'll hop aboard a space shuttle to get the view we need.

As we circle above the equator, **infrared** sensors describe the landscape below. This shows us where the trees are. The sensors also record temperature variations. This tells us how much forest is on fire.

We study the monitors. Dark green areas are heavily forested. Light green areas have been logged for lumber or fuel wood. After the trees were removed, the land was planted with crops like tea, bananas, and coffee. Other land became cattle pastures.

Now we compare the size of these areas to how big they used to be. This tells us how much rain forest has been lost. What we see makes us stop and think.

Between 15 and 20 percent of the Amazon rain forest is gone. Thirty thousand squares miles disappear every year, due to logging, **slash burning**, and drought. Beautiful Costa Rica has lost 25 percent of its forest. And on the Indonesian island of Borneo, 80 percent has vanished.

Drought, or dry El Niño years, makes everything worse. When rain forest turns dry, controlled burns often get out of hand. Smoke from the fires kills seedlings. It also causes health problems for people and animals.

El Niño is a warm current in the Pacific Ocean that flows southward along the west coast of South America. This warm current usually appears every year from December to March. Scientists often use the term to describe a longer event that has widespread effects. They believe El Niño is related to a shift in air movements over the tropical waters. These changes disrupt air movements too. Severe weather conditions can develop, such as droughts, violent storms, and destructive floods.

Heavy smoke from forest fires rises into the sky. Sometimes it gets so thick that moisture droplets can't get big enough to become rain. Rainfall shuts down completely. So drought, fire, and smoke change the climate by creating more drought, fire, and smoke.

The Borneo rain forest is unique. Most of the canopy trees are dipterocarps (dip-TARE-oh-karps). These trees flower *only* during El Niño years. And *every* dipterocarp tree flowers, fruits, and drops its seeds in the same six-week period. Dipterocarp seeds cover the forest floor like a crunchy, brown carpet.

Then the feast begins! Parakeets and jungle fowl swoop in. Wild boars come running. Hungriest of all are Borneo's orangutans. These apes live only in Borneo's rain forest. The animals stuff themselves. And rain forest people gather seeds to sell. Still, there are enough left over to grow into seedlings.

But a decade of uncontrolled logging is changing that. Chain saws drop dipterocarps for lumber. When dry weather comes, fewer seeds fall. Hungry animals overrun the areas where logging is off-limits. The animals can't find enough to eat. So there are no seeds left over to sprout and grow.

In the last ten years, the Borneo orangutan population has dropped from 15,000 to 7,000.

Indonesia is a poor country. Timber sales bring in some 8 billion dollars every year. But that will change, too, if seedlings cannot sprout. As forest disappears and partnerships are lost, everyone is hurt.

In Costa Rica, cattle ranchers have claimed most of the vanished rain forest. In 1950, pastureland covered about $1/8$ of the country. Today, pastures cover more than $1/3$ of the land. But much of it has been abandoned. It no longer grows grass that cattle can eat because of poor land use. When that happens, ranchers simply clear more forest.

All that might not be so bad if the beef fed the Costa Rican people. But most of it is sent out of the country, or *exported*, especially to America. Costa Rican beef is much less expensive than the American-grown kind. It's the same story in Nicaragua, Guatemala, and El Salvador. While native peoples often go hungry, Big Macs are eaten in other parts of the world.

Passing back over Brazil, our sensors pick up the Transamazon Highway and the damage it has created. The road was supposed to be Brazil's "solution for 2001." The idea was to open the heart of Amazonia to would-be farmers.

A 12-mile-wide strip along the highway was cleared for agriculture. The government of Brazil expected 100,000 families to buy 250-acre farms. It didn't happen that way. Some 7,500 families gave it a go. But crop failures and malaria chased most of them home again.

When large tracts of rain forest are cleared, malaria epidemics hit those areas. Why?

Mosquitoes carrying the sickness usually stay in the canopies. They feed on the animals living in the treetops. When trees fall, the mosquitoes move lower to feed. And that's where they find people.

The highway and farms fragmented the forest. Trees near the edge of forest fragments die more easily than those in the middle. The main cause is the wind, which blows harder in open spaces. The problem is worse during drought years. The trees are already weak. Put all that together. Then add fire and you get total destruction.

Infrared images show us thousands of acres of burning rain forest. Half are wildfires. In wet years, they would never get out of control. But recent dry years have given the trees just enough moisture to survive. They were ready to go up in flames.

The rest of the fires were set to clear forest for agriculture. These fires are part of the slash-and-burn method used by outsiders. Large plots of land are cleared. The felled trees are then burned. The ash contains some nutrients for the seeds. But when the rain comes, those nutrients wash into streams and rivers. They finally flow out to sea.

Within two or three seasons, the soil no longer grows crops. More land is cleared, and the wasteful cycle begins again.

After a couple of seasons, the new plot won't grow crops. Often, outsiders then return to the original plots. They recut any new growth and burn the slash.

Recutting too soon makes the land completely infertile. The forest cannot recover and rebuild. That leads to **erosion**.

In uncut forest, less than 1 ton of topsoil per $2^1/_2$ acres is washed away each year. But when crops are planted on that same piece of land, 20 to 160 tons wash away. And in pastures, up to 300 tons a year wash away.

Native farmers have a better way to clear the land. Their method is to clear a small plot. Then the debris is left to decay. The litter slowly releases nutrients and shades the early crops. After several growing seasons, the farmer leaves the land alone. During long **fallow** periods, the forest rebuilds itself.

This "primitive" conservation method works for outsiders too. In 1951, Quaker settlers bought a large piece of Costa Rican rain forest. They named it Monteverde and started a dairy farm.

Most of the land had never been logged. But a few small patches had been slashed-and-burned for gardens. The Quakers tested the soil. The uncut land was rich enough for pasture. But where it had been burned, it was very poor. The Monteverde settlers decided they would never slash-and-burn.

They cleared the land with axes. They later used chain saws for the bigger trees.

First, they removed the undergrowth and planted grass seed. They left the tall trees to shade the baby grass. When the grass matured, they cut some trees and left them to rot. The debris returned nutrients to the soil. Other trees were left for shade and to hold the topsoil in place.

Today, cattle graze between fallen trees, eating healthy grass. The Quakers change pastures every day. That way, cattle can't overgraze and the soil stays healthy.

Like the natives, the Quakers respect the fragile forest. But other outsiders would rather use it up. And if the rain forest vanishes, we all lose.

CHAPTER 8

Nature's Mega-Supermarket

Have you ever wandered around one of those mega-supermarkets? You know, those places where they sell everything from ice cream to jeans to tires for your car? To find a certain item, you might have to ask the guy who stocks the shelves.

Well, the rain forest is nature's mega-supermarket. And the people who know its "shelves" best are the natives who live there. Let's wander through nature's mega-market.

Hungry? In the produce section, we find fruit, seeds, berries, nuts, and teas. Not to mention some 80,000 **edible** plants. The butcher shop offers fish, game birds, and meat. There is pork from wild pigs and beef from wild cattle.

Wild cattle are resistant to heat and disease. In America and elsewhere, they are crossbred with domestic breeds. This makes domestic cattle stronger.

Maybe you're in the mood for sweets. The grocery department has sugar, honey, juice, spices, and cacao (cocoa). What wonderful desserts we could whip up with those!

Is chocolate endangered? Scientists think so. Cacao trees have been hit hard by insects and fungi. And people are eating more chocolate than ever before.

Researcher Jeff Moats may have an answer. The ground seeds of the cupuacu (koo-poo-ah-SOO) fruit taste very much like chocolate. Moats is making cupuacu candy bars. A percentage of his sales will go toward protecting the rain forest.

The rain forest superstore has a whole lot more than food. There are plants from which clothing can be made. Rattan is used for furniture. There is wood for pulp and paper and jute for string and rope. Fuel, oil, varnish, resin, inks, and dyes can be found. Ingredients for insecticides, soaps, perfumes, polishes, and makeup are also found here. All this and more, just for the taking, can be found in the rain forest.

Let's not forget the pharmacy. You thought all drugs were cooked up in test tubes? Many are. But half the medicines we use come from plants. And 25 percent of all prescription drugs come from tropical flora. Many of these "wonder drugs" have been used for centuries by rain forest people.

Many plants contain alkaloids. These are colorless, often bitter chemical compounds. Their icky taste keeps insects and other animals from eating the plants.

Tropical plants are twice as likely to have alkaloids as temperate plants.

Alkaloids are the base for many medicines. They are used as blood pressure boosters, fever reducers, painkillers, muscle relaxants, and cancer fighters.

The rosy periwinkle from Madagascar produces 75 different alkaloids. Two are proven cancer fighters. They are 99 percent effective against leukemia and 80 percent effective against Hodgkins disease. Before scientists found them, Hodgkins patients had only a 19 percent chance of survival.

When Europeans settled in the Amazon in the 1500s, many fell victim to malaria. But the natives had a cure. It was fever bark. It came from the cinchona (sin-CHOH-na) tree.

The bark's active ingredient is quinine. Quinine is still used to fight malaria, headaches, and infections like pneumonia.

Other settlers fell victim to Indian arrows. The arrows didn't kill them, but their toxic tips did. The poison was curare (que-RAR-ee). It comes from a combination of plants. Curare works by shutting down the nervous system and relaxing the muscles. When the **diaphragm** muscle relaxes, a person stops breathing.

Today, curare puts people to sleep in hospital operating rooms. It is used as an anesthesia. It is also used to treat polio, lockjaw, and epilepsy.

The science of plant and people partnerships is called *ethnobotany*. Ethnobotanists often live with native peoples to learn which plants they use. They eat the same foods, sleep in the same huts, and wear the same simple clothing.

By observing traditional healers, ethnobotanists have made many valuable medical discoveries.

Homolanthus nutans isn't the biggest rain forest tree. But its healing properties are amazing. Samoan healers have long used its bark to treat hepatitis. And now it may be used to treat AIDS. America's National Cancer Institute (NCI) has found a compound in the bark that looks promising. It's called prostratin (pro-STRAY-tin).

The discovery may help save the Samoan rain forest. Logging has already destroyed 80 percent of the tropical forest in Western Samoa. But if prostratin really does fight AIDS, the NCI will donate half the profit to save the forest. That could be millions of dollars.

Let's look at other solutions to rain forest destruction.

CHAPTER 9

SOLUTIONS

One simple fact of life on Earth cannot be ignored. Animals, including people, breathe in oxygen and breathe out carbon dioxide (CO_2). Plants take in CO_2 and release oxygen. This basic partnership has worked well for millions of years. So think what could happen if plants disappear! Where would we get oxygen to breathe?

Now think about what happens when plants burn. Plants take in CO_2 and store it. When they burn, the smoke is full of it. It rises into the atmosphere.

CO_2 is a "greenhouse gas." It holds heat. In the atmosphere, it traps heat that is trying to escape from the surface of the earth. This is the "greenhouse effect" you've heard so much about. Our planet is getting warmer. The cause is greenhouse gas in the atmosphere.

Earth has been warming steadily for 30 years. If the trend continues, a number of things could happen.

The polar ice caps could melt. If that happens, sea levels will rise. Coastal regions will flood. That might not bother you if you live in Kansas. But if you live in Florida, Portland, Los Angeles, or Washington, D.C., it certainly would.

And if you live in Kansas, you wouldn't be happy, either. Warmer temperatures will change wind patterns. You'll get less rain. Drought will kill your crops. You may even find yourself living in a desert.

Industry is a major source of greenhouse gas. Many countries are trying to reduce industrial emissions. But the process is slow. A quicker fix would be to control forest burning. The burning of trees produces 35 percent of the greenhouse gas going into the atmosphere.

Remember how smoke in the atmosphere reduces rainfall? Less rain means more trees will die and forests will burn easier.

More forest fires = more smoke = less rain = even fewer trees.

Trees and animals die. Partnerships are lost. Species, some still undiscovered, disappear forever. Somewhere in the world, this is happening right now.

Only 10 percent of all tropical plants have been tested for medicinal qualities. Many kinds only grow in certain small areas. When those areas burn, their plants could vanish. Suppose the cure for cancer or AIDS goes up in smoke. Suppose it already has.

There are solutions. Many rain forest regions simply aren't suited to agriculture. These areas should be set aside and protected at all costs. So must the creatures that live within their boundaries. And this should be more than "paper" protection. Laws and treaties need to be enforced with action.

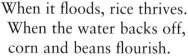

Other areas can sustain agriculture. The Amazon's fertile floodplains are a good example. The river regularly deposits silt and debris, rich with nutrients. This soil can grow three crops a year and stay fertile.

When it floods, rice thrives. When the water backs off, corn and beans flourish.

After a natural burn, the forest rebuilds in a pattern. First come herbs, then fast growing trees. Finally, canopy trees. Cleared land can be planted in a similar pattern. Pineapple, sugar cane, and beans are planted first. These quick crops are grown for one year. The second year, papaya or bananas are planted. These may produce for five to ten years.

Meanwhile, slow growing trees like the peach palm are planted. Both the fruit and palm hearts are staples in the native diet. These trees start producing after eight years and grow for 50 years.

After that, the land must go fallow. Tests show it should stay unused for at least 14 years. In that time, the soil can rebuild.

Slash-and-burn needs to stop. Better logging methods can be used.

One way is to cut the lianas that hold tree crowns together. When held together with lianas, trees that don't produce lumber fall with those that do.

Logs should be removed by helicopter. That saves trees from being cut down to make roads for hauling lumber.

All this, of course, is more expensive in the short run. But in the long run, it is worth every extra penny. And there are ways to offset some of the costs.

Some countries offer tax benefits to companies that practice conservation. Others sell carbon bonds. These are like savings bonds. But instead of earning interest, they earn rain forest protection.

Carbon bonds are "certified tradable offsets," or CTOs. Countries such as Costa Rica, Brazil, and Panama allow industries to release greenhouse gases if they purchase CTOs. The money from CTOs buys and protects tracts of rain forest, which absorbs some of the carbon released into the air by industry.

Do conservation programs work?

"Some do, and some don't," said Sabrena Rodriguez of the Rainforest Alliance. "There are no one-size-fits-all answers. The conservation community seeks solutions to problems one village at a time. We must constantly test new possibilities and look for new solutions."

Sabrena said kids can help by encouraging their parents to make wiser buying decisions. "Americans consume a large percentage of tropical wood and agricultural products. Our buying decisions have a strong impact on what growers, foresters, and governments choose to do."

Groups like the Rainforest Alliance work within the international community to protect the rain forest.

To find out more, and learn more ways you can help, visit their web site: http://www.ra.org